OCEAN

NATIONAL GEOGRAPHIC NATURE LIBRARY

by Patricia Daniels

NATIONAL GEOGRAPHIC SOCIETY

Washington, D.C.

OCEAN

seahorse

NATIONAL GEOGRAPHIC NATURE LIBRARY

*Waves pound the rocky coast of
Hawaii Volcanoes National Park.*

Table of Contents

WHAT IS THE OCEAN? 6

Under the Ocean 8

Moving Waters 10

Red Sea

1 Ocean Zones 12

The Food Web 14

manatee

2 Coasts 16

Between the Tides 18

Kelp Forests 20

Glorious Mud 22

Coral Reefs 24

lionfish

3 The Sunlit Waters 26

Seabirds 28

Marine Mammals 30

Life in Motion 32

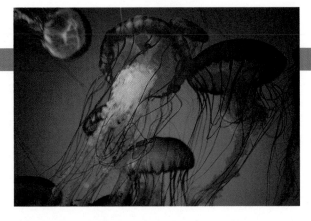

jellyfish

4 **The Open Ocean** 34
The Ocean Floor 36
Vents 38

anglerfish

5 **The Ocean Tour** 40
Pacific Ocean 42
Atlantic Ocean 44
Indian Ocean 46
Arctic Ocean 48
Antarctic Waters 50

leafy sea dragon

6 **The Human Factor** 52
Exploration Today 54

submersible Alvin

Did You Know... 56
Glossary 58
Index 59   Credits 60

WHAT IS THE OCEAN?

Earth is a water planet. It is wrapped in an ocean, a huge body of salt water that covers three-quarters of the Earth's surface. Although we give different names, such as Atlantic and Pacific, to various parts of the ocean, it is one single body of water. It surrounds Earth's major landmasses, the seven continents. North America, South America, Asia, Africa, Europe, Antarctica, and Australia are all surrounded by ocean.

The ocean is its own world, with mountains, valleys, volcanoes, and a wealth of animals from tiny to huge. Here are some things we know about the ocean:

- The ocean formed about 4½ billion years ago.
- The ocean is 96.5 percent WATER and 3.5 percent DISSOLVED SALT.
- Ocean water stores HEAT. It keeps the world from getting too cold.
- The ocean holds almost all LIVING MATTER on Earth—97 percent.
- Most of the OXYGEN we breathe comes from plants in the ocean.

ray

black smokers

deep-sea anglerfish

tube worms

butterflyfish

sea star

dolphins

butterflyfish

angelfish

coral

Under the Ocean

Most of us have seen the ocean only at its surface. Dive down into the water, and another world appears. Each of the continents continues underwater in a shallow shelf. About 600 feet down, the land slopes more sharply to the ocean floor. There, stretching across thousands of miles, are tall mountains, valleys, volcanoes, and flat plains—much like the land above, but mostly unexplored.

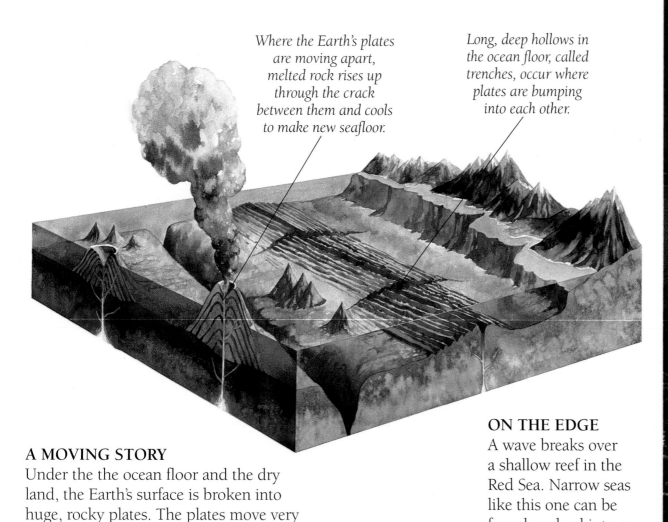

Where the Earth's plates are moving apart, melted rock rises up through the crack between them and cools to make new seafloor.

Long, deep hollows in the ocean floor, called trenches, occur where plates are bumping into each other.

A MOVING STORY
Under the the ocean floor and the dry land, the Earth's surface is broken into huge, rocky plates. The plates move very slowly, pulling apart in some places and crunching together in others.

ON THE EDGE
A wave breaks over a shallow reef in the Red Sea. Narrow seas like this one can be found wedged into an opening in a continent or squeezed between the land and the deep ocean basin.

8

Moving Waters

Just as there are rivers on land, there are rivers in the ocean. They are called currents—great streams of water that flow within the ocean. Some currents are warm, and some are cold. Steady winds create currents on the water's surface. Below it, currents are caused by differences in the temperature and saltiness of water. Curents help even out the world's temperatures by mixing warm and cold water.

Benjamin Franklin was the first person to map the Gulf Stream.

WHAT A DRAG

Winds that always blow in the same direction—called prevailing winds—drag on the ocean's surface to make surface currents. A 50-mile-an-hour wind creates a 1-mile-an-hour current.

OCEAN RIVER ▶

The Gulf Stream is a strong ocean current that flows up the East Coast of the United States and out into the North Atlantic Ocean. This satellite image shows the warm waters of the current in orange.

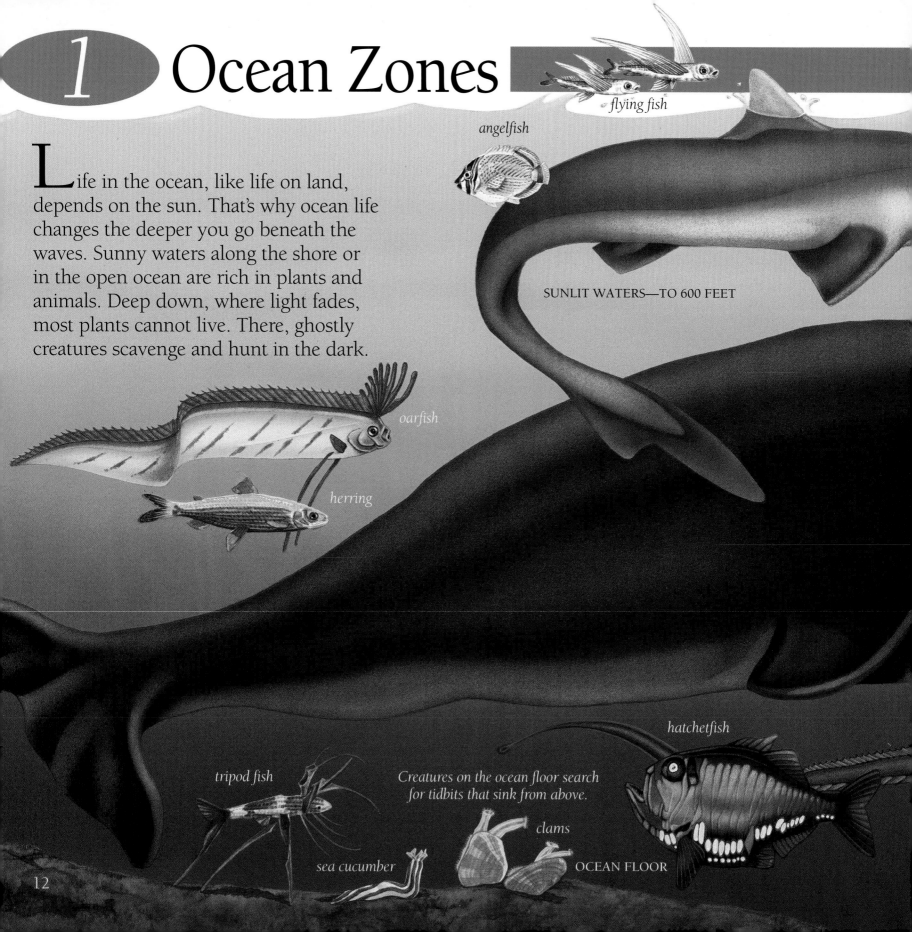

1 Ocean Zones

flying fish

angelfish

Life in the ocean, like life on land, depends on the sun. That's why ocean life changes the deeper you go beneath the waves. Sunny waters along the shore or in the open ocean are rich in plants and animals. Deep down, where light fades, most plants cannot live. There, ghostly creatures scavenge and hunt in the dark.

SUNLIT WATERS—TO 600 FEET

oarfish

herring

hatchetfish

tripod fish

Creatures on the ocean floor search for tidbits that sink from above.

clams

sea cucumber

OCEAN FLOOR

12

SHORE

plankton
(magnified)

jellyfish

hermit crab

lobster

shark

sea lion

The sunny, food-filled waters of the shore support many plants and animals.

sea turtle

Fish, seabirds, and tiny plants and animals thrive in the ocean's bright upper waters.

sea anemone

whale

TWILIGHT ZONE—600 TO 3,000 FEET

sea star

manta ray

LAYERS OF LIFE
Plants and animals live in their own zones, or layers, of the ocean. The sunlit zone, the top 300 feet of water, is the brightest. About 1,000 feet below the surface, light is dim. Only a few plants can survive. Below this twilight level, waters are dark and cold. Food is scarce, and animals are rare.

lanternfish

In the dim waters hundreds of feet below the surface, animals eat each other and bits of food that sink from above.

gulper eel

deep-sea
anglerfish

brittle star

THE DEEPS—3,000-36,000 FEET

Animals that live in the deep ocean often have strangely shaped bodies.

The Food Web

The most important web in the ocean has nothing to do with spiders—except spider crabs. It is the vast food web that connects all ocean life. Using energy from the sun, plants make food. Animals eat plants and other animals. Other creatures, including tiny one-celled organisms called bacteria, eat and recycle the leftovers.

FILTER AND SCRAPE
Some ocean creatures don't have to speed to find food. Prickly sea urchins move slowly as they scrape their meals off rocks. Anchored to rocks or the seafloor, barnacles don't move at all. They filter food from the water.

VEGGIE TIME
Plants are the most abundant source of food in the ocean. That's a good thing, because it takes a lot of them to support just one animal. This manatee, grazing on underwater plants in the Florida Keys, can eat up to a hundred pounds of plants a day.

One strand of the food web connects sunlight to sharks. Almost all food webs start with the sun. Thousands of tiny organisms called plankton turn sunlight into food. Those plankton feed hundreds of fish. One hungry shark eats all those fish.

PICNIC ON THE WAVES
A sea otter uses its stomach as a plate while munching on a crab. Many sea creatures feed directly on other animals.

GULP!
Sometimes the biggest creatures eat the smallest ones. Blue whales, the largest animals on Earth, eat various tiny, floating animals called krill. The whales scoop up krill, along with up to 55 tons of seawater in their huge mouths.

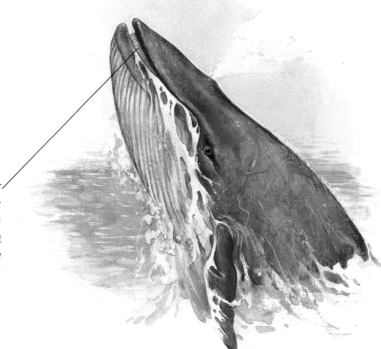

The blue whale, like some other whales, has a horny, comb-like substance called baleen (buh-LEAN) in its mouth. It uses the baleen like a strainer to sift krill and other tiny organisms from seawater.

15

Many poisonous animals live in coral reefs, including Australia's deadly blue-ringed octopus.

Ocean life thrives in the shallow waters where the land meets the sea. Coasts give animals and plants sunlight, shelter, and food. Yet coasts can also be tricky homes because they are changeable. Waves break constantly. At low tide, water falls away from the shore. At high tide, water rushes in. Tides pull water in and out, exposing and covering the land at the water's edge. Life adjusts to this ebb-and-flow existence along the many kinds of coasts.

ROCKY REST STOP
A harbor seal basks on a rock on the Pacific Northwest coast. Harbor seals often live along rocky coasts like this one, where they can find fish and crabs.

GROWING SHORE ▶
This shore is alive. Tiny animals called coral polyps (POL-ehps) build stony skeletons as they grow. Branching staghorn coral builds a reef underwater. Over time the coral may add shoreline to the coral island in the background.

DANDY SAND

The grains of sand along this beach in Maui, Hawaii, were bigger once. Waves grind down rocks, shells, pieces of coral, and even chunks of lava to make a sandy shore. Animals often survive in the sand by burrowing.

MUD AND MANGROVES

The stilt-like roots of mangrove trees dig into the mud along a warm, quiet shore, creating a rich habitat for small creatures. Land, fresh water, and ocean animals mix in these sheltered waters.

17

Between the Tides

Tides move water in and out all day, baring and covering parts of the shore. A tidal shore is a series of little worlds, each with its own environment. The highest area is just reached by waves. The mid-shore level is covered by water for hours each day. The lowest part is almost always underwater.

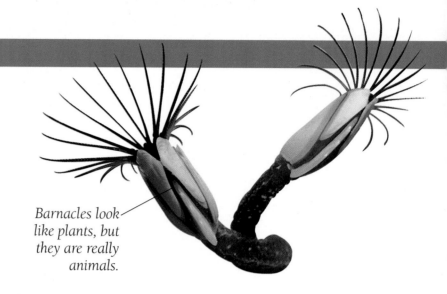

Barnacles look like plants, but they are really animals.

EAT WITH YOUR FEET
Goose barnacles attach to rocks by their heads. When they are underwater, they kick food into their mouths with feathery structures called cilia (SIH-lee-uh). When the tide goes out, barnacles retreat into hard coverings.

Nudibranchs (NOO-duh-branks), or sea slugs, like to stay in shallow waters where they hunt for creatures such as sea anemones.

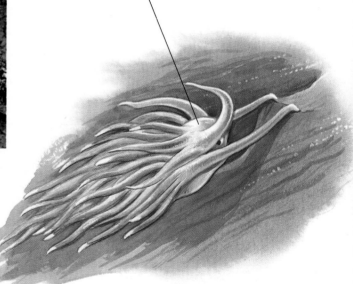

AT HOME IN THE TIDES
A hermit crab braces itself in the water of a tidal pool. These little crabs are typical shore animals. They can live in water or land and eat almost anything.

STARRY WATERS

Like colorful toys, sea stars, snails, and other creatures spread across a rocky pool in British Columbia, Canada. These animals prefer to live in the mid-shore level of a tidal shore.

UPS AND DOWNS

Tides bring drastic changes to some coasts. Waters rise and fall by 40 to 50 feet each day in Canada's Bay of Fundy.

19

Kelp Forests

Kelp plants are the largest plants in the ocean. These huge seaweeds form tall underwater forests near cold, rocky shores. Growing as much as a foot a day, a kelp plant reaches 130 feet in length. Like a forest on land, kelp supports different kinds of life from its sunlit top, or canopy, to its shadowy bottom. Kelpfish, kelp snails, sea urchins, and many other creatures form a kelp community.

FLOATING FRONDS
A harbor seal dives into the dim lower reaches of a kelp forest. Kelp stays upright with the help of gas-filled floats in its leaf-like fronds. Root-like structures called holdfasts help the seaweed cling to the rocky bottom of the ocean.

IT'S A JUNGLE DOWN THERE

Like explorers in a rain forest, two divers make their way through giant kelp plants near Baja California, Mexico. Giant kelp is found in cold waters along many coasts. It cannot survive in warm water.

SEAWEED ICE CREAM?

Giant kelp produces a starchy chemical called algin (AL-jihn) that is used in products from tires to lacy fabrics. It is also added to many foods, making smoother, thicker ice cream, pies, and frostings.

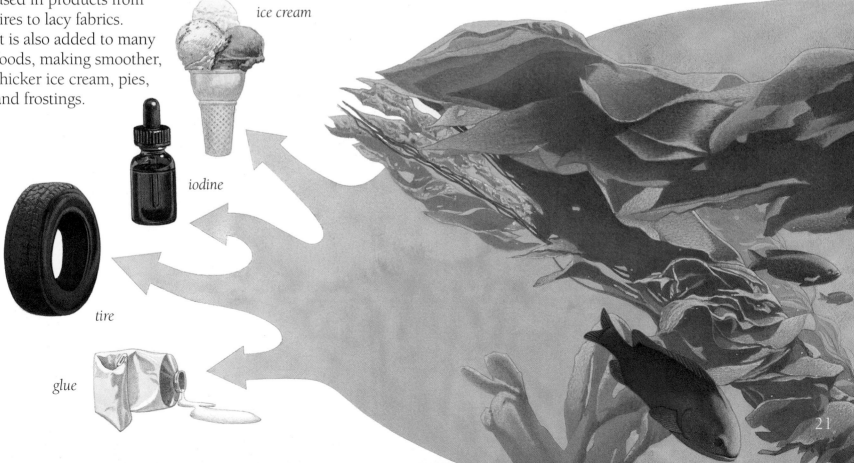

ice cream

iodine

tire

glue

21

Glorious Mud

For many shore animals, mud is heavenly. The muddy bottoms of swamps, marshes, and estuaries—areas where salty water and fresh water mix—form in quiet areas where fine soil settles after being carried by waves or rivers. Crabs and snails do well here, climbing among grasses and tree roots. Clams, worms, and sea cucumbers burrow into the soft bottom. Many of the small animals feed on the rich soup of detritus (dih-TRY-tuhs)—floating bits of plants and animals—that sink into the mud.

Fiddler crabs aren't fussy. They pick their food right out of the mud.

THE FISH THAT WALKED

The mudskipper is a fish that refuses to act like a fish. Known also as the "climbing perch," it uses its fins as legs to climb out of the water along warm, muddy shores. It may even climb trees to hunt for insects and crabs.

EGG TIME

Soft and sticky is fine with the horseshoe crab. Each summer females scoop out holes in mud or sand in which to lay their eggs. They need to watch out, though. Seabirds like to come by for an egg feast.

The horseshoe crab is not a crab at all. It is a member of an ancient group of animals related to spiders.

BRING YOUR OWN SPOON
Roseate spoonbills know how to dig for a meal. Striding along marshy bottoms, they sweep their rounded bills from side to side through the water, picking up small fish and crabs.

THEY'RE ROOTING FOR YOU
Curving, stilt-like roots of mangrove trees make their own special habitat in tropical mangrove swamps. The roots trap food and soils, attracting a mixture of land and sea creatures that live in and around them.

Coral Reefs

Some coasts are made of rock, some of sand, and some of mud. Coral coasts are made of thousands of skeletons that form a coral reef. Young coral creatures, called polyps begin their lives as tiny swimming specks. Then a polyp attaches itself to a hard surface such as a rock. There it grows and forms its own external skeleton of solid limestone. Some corals live in colonies, while others live alone. All are delicate, able to live only in clean warm water.

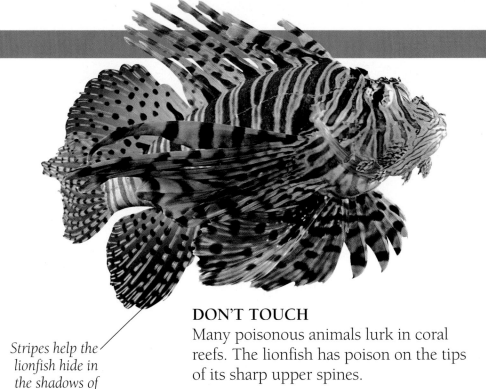

Stripes help the lionfish hide in the shadows of a coral reef.

DON'T TOUCH
Many poisonous animals lurk in coral reefs. The lionfish has poison on the tips of its sharp upper spines.

Coral can grow as much as one inch every year.

CORAL COTTAGE
Coral polyps build their skeletons from substances pulled from seawater. A coral has tentacles it extends to collect food and bring it to its mouth.

CORAL CITY
Corals come in many shapes, from branching to round, and in almost every color. A coral reef is like a city with various neighborhoods for sea plants and animals.

REEF ATTACK

A barracuda lunges out of its hiding place in a coral reef, scattering a school of silversides. Sharks, barracudas, and other predators often cruise through and around coral reefs. Worldwide, coral reefs feed and shelter at least a million species of animals and plants.

The crown-of-thorns sea star eats coral by turning its stomach inside out and digesting coral tissues.

THREATS

Coral reefs are threatened by many things, including human activities that disrupt a reef's balance of plants and animals. One natural enemy is the hungry crown-of-thorns sea star.

25

3 The Sunlit Waters

The upper waters of the open ocean, the top 600 feet or so, are like a floating meadow. Huge numbers of tiny plants and animals called plankton drift through the sunlit waters. Many creatures feed on the plankton. The plankton-eaters are hunted by sharks and other predators.

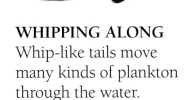

The whip-like tail of this plankton is called a flagellum (fluh-JEL-em).

WHIPPING ALONG
Whip-like tails move many kinds of plankton through the water.

ON THE MOVE
A green sea turtle heads out to sea. Sea turtles are among the ocean's great travelers. They swim thousands of miles before returning to beaches to lay their eggs.

About 60 percent of the sun's energy is absorbed by the top three feet of seawater.

CROWDS OF KRILL
Krill, small shrimp-like plankton, swim together by the billions. They are valuable food for many whales and fish.

SCHOOL DAYS ▶
Circling through the sunlit sea, barracuda seek out prey in the Pacific. Ocean fish often collect in groups, called schools, to graze on plants and to hunt.

Seabirds

Seabirds are the living link between the worlds of land and ocean. Although no bird lives its entire life at sea, many birds find their food there. Albatross and other open ocean birds are far-fliers that may stay out to sea for years. Other birds, such as pelicans and cormorants, dive for fish from the shore. Penguins don't fly at all, but they do swim with grace and speed.

THE PELICAN'S POUNCE
Like a fishnet with wings, a diving pelican scoops up its prey. Pelicans are well adapted for sea life. Their webbed feet move them strongly through the water, while their stretchy throat pouches can swell dramatically to hold fish.

Storm-petrels fly close to the water, hopping across the tops of waves.

CHICKEN OF THE SEA
Wilson's storm-petrels sometimes follow oceangoing ships, picking up scraps as farm birds often do in barnyards.

LOOKING FOR FOOD—ANY FOOD
Gulls are the most common seabirds and the most familiar ones to humans. These noisy, pushy birds will hunt fish, but they feed on garbage as well. Gulls are superb and acrobatic fliers.

SHORE LEAVE
When it comes time to start a family, an albatross lays a single white egg on land. The rest of the time, the albatross is the ruler of the ocean winds. With a wingspan up to 12 feet wide, an albatross can soar for thousands of miles on strong sea breezes. Like many ocean birds, the albatross drinks salt water if fresh water can't be found.

29

Marine Mammals

Some of your relatives live in the ocean. Like you, they are mammals, warm-blooded animals that grow hair and give birth to live young. Marine mammals are well adapted to ocean life. Their shapes are streamlined for swimming, and almost all of them can stay under for a long time.

SEAL APPEAL
A gray seal checks out the surface scene. A layer of fat, called blubber, under a seal's skin helps it stay warm in cold water.

DOLPHINS IN THE FAMILY
Dolphins, such as this one, belong to a group called cetaceans (sih-TAY-shuns), which includes porpoises and whales.

Pakicetus (pak-ih-SEE-tuhs), a whale ancestor, lived on riverbanks about 50 million years ago.

A WHALE OF A BABY
A humpback calf follows its mother. Whales and other cetaceans give birth underwater. Their babies swim right away.

LAND WHALE
Whales, the biggest creatures on Earth, evolved long ago from a land creature that looked like a sharp-snouted fox.

IN YOUR FACE ▶
The round forehead of a beluga whale holds a fatty structure called a "melon."

Life in Motion

There's no place to rest in the sunlit upper zone of the open ocean. Plants and animals there are always on the move—floating or swimming. Huge masses of plankton drift and wiggle through the sunny waters. Fish, squid, and marine mammals hunt the plankton—and each other.

Sailfish vs. grouper? No contest. The slender, compact sailfish is the fastest fish in the sea. It can swim nearly 70 miles an hour.

DANGEROUS WATERS
Sharks are found throughout the ocean. Most, such as the reef sharks above, are deadly predators, but a few eat only plankton.

LONG-DISTANCE CHAMPIONS
Tuna display a torpedo shape that is typical of fish in the upper ocean zone. Smooth and streamlined, they are designed for speed. Tuna also have great endurance. Some have been known to swim the width of the Pacific Ocean.

IT'S HARD TO EAT JUST ONE
By swimming in a school, these glassy sweeper fish may confuse predators trying to pick out one prey. Their color helps the fish blend into the shimmering background.

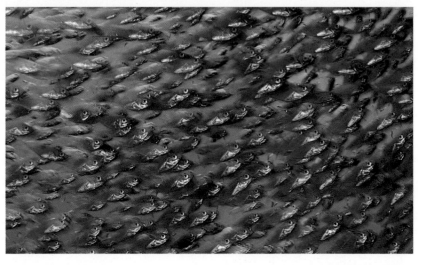

FLOATERS
More than 400 species of jellyfish float and swim in the open sea. Their bell-shaped bodies help to keep them afloat. Some jellyfish, such as sea nettles, can deliver a painful sting.

A s you dive below 600 feet, light dims. At 1,000 feet you enter complete darkness. Normal plants cannot grow here, so plant-eating fish rely on tidbits that sink from above. Because food is scarce, animals in the deep sea are few and far between—but what they lack in quantity, they make up in strangeness. Their distorted, glowing bodies are adaptations to the hide-and-seek life of the deep ocean.

OPEN MOUTH, INSERT FISH
The huge jaws of this viperfish allow it to swallow big fish in a region where food is hard to find.

Organisms called bacteria live within the tissue that the deep-sea anglerfish uses as bait. Only the female anglerfish has this glowing lure.

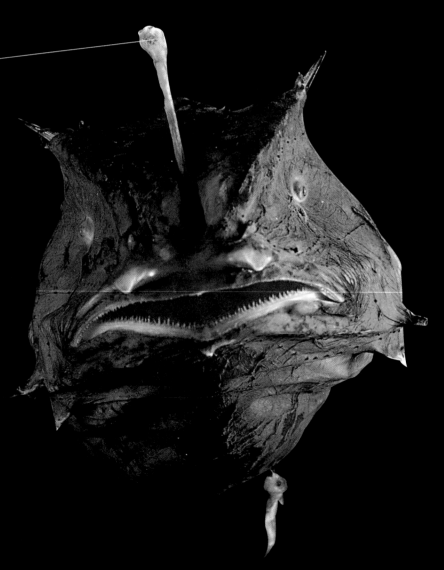

COME AND GET IT!
It may not look yummy to you, but some deep-sea fish are attracted to the fleshy, glowing "bait" on the anglerfish's front spine.

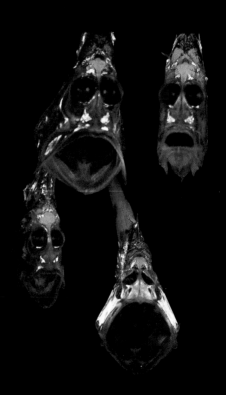

WHAT BIG EYES YOU HAVE!
The bulging eyes of these hatchetfish look upward, searching for glowing prey.

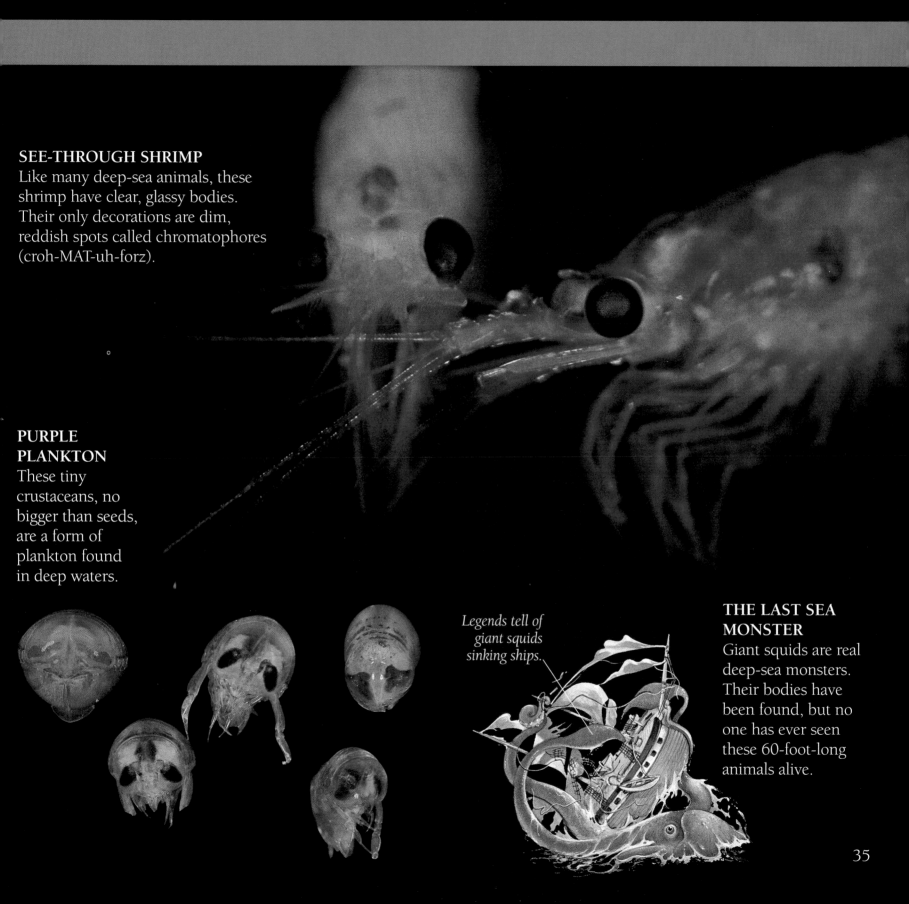

SEE-THROUGH SHRIMP
Like many deep-sea animals, these shrimp have clear, glassy bodies. Their only decorations are dim, reddish spots called chromatophores (croh-MAT-uh-forz).

PURPLE PLANKTON
These tiny crustaceans, no bigger than seeds, are a form of plankton found in deep waters.

Legends tell of giant squids sinking ships.

THE LAST SEA MONSTER
Giant squids are real deep-sea monsters. Their bodies have been found, but no one has ever seen these 60-foot-long animals alive.

The Ocean Floor

The ocean floor varies in depth. Near shore it can be just under the water's surface. In the open ocean, the bottom lies miles below the waves. The deep sea bottom is another world. It has no seasons, no night or day. The temperature stays a few degrees above freezing. The few animals that live here feed by eating bits of food that sink from above.

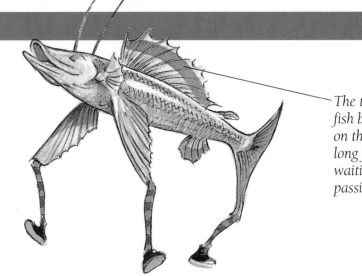

The tripod fish balances on three long fins, waiting for passing food.

Curving mounds of lava that have cooled quickly in the water are called pillow lava.

UNDERSEA PILLOW
Where plates in the Earth's crust pull apart under the ocean floor, lava wells up.

LURKING IN THE DARK
A scorpionfish waits on the ocean floor for an unsuspecting victim to swim by. Scorpionfish have spines that deliver painful stings.

CURIOUS CUTLASS
A silvery cutlass fish noses a tube anemone on the seafloor. Cutlass fish live in the twilight zone where waters are not totally dark.

LIVING BASKET

Clinging to a handy rock, a basket star waits to snatch plankton from the current. Like the cutlass fish, the basket star lives in the dim zone where small amounts of light still reach.

Large spider crabs weigh more than 40 pounds.

KING-SIZE CRAB

The giant spider crab can span 12 feet from one leg tip to the other.

Vents

In the late 1970s, scientists discovered a new form of life—in the last place they expected. On the deep ocean floor, along the edges of Earth's plates, they found very odd creatures. Giant tube worms, huge clams, and blind lobsters crowded around openings, called vents, in the Earth's crust that released scalding hot water. The mineral-rich water supported a new kind of bacteria that could make energy from chemicals. The bacteria fed rich life near the vents.

BLACK SMOKERS
Cooked by hot rocks in the Earth's crust, burning hot water shoots out of vents. Sulfur and other minerals in the water settle around the vents, building up into chimneys called black smokers.

The name black smokers comes from the dark water, colored by sulfur, that rises from their chimneys. The water is also darkened by clouds of bacteria.

Water from the vents can be as hot as 720°F.

GUTLESS WONDERS
Giant tube worms that live in thick clusters around vents can grow up to ten feet long. They have no digestive organs. Bacteria living inside them use vent chemicals to make food that the worms can use.

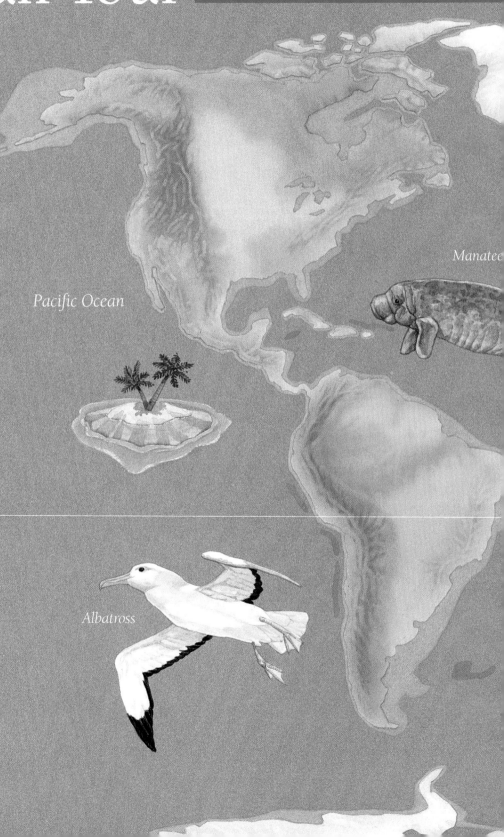

Although the ocean is one single body of water, people have given different names to different parts. The Pacific Ocean is the largest part, covering one-third of the Earth. Next is the Atlantic, with its busy shipping lanes and rich fishing grounds. The Indian Ocean, third biggest, is known for its changeable currents and powerful winds. The Arctic Ocean, the smallest part, is covered by rafts of shifting ice. Scientists do not consider the waters around the Antarctic a distinct part of the Earth's ocean.

Manatee

Pacific Ocean

Albatross

WATER, WATER EVERYWHERE
The continents of the world divide the ocean into its named parts. The southern half of the Earth is almost entirely water, centered on the continent of Antarctica.

Arctic Ocean

Polar bear

Gray whale

Atlantic Ocean

Indian Ocean

Great white shark

Rockhopper penguin

Pacific Ocean

The word "pacific" means peaceful, but the Pacific Ocean is far from calm. Active volcanoes rim the Pacific's shores and rise from its floor. The biggest and deepest ocean, the Pacific holds more than half the world's unfrozen water. It also has the world's deepest spot—the Mariana Trench, 35,827 feet below sea level.

HOT TIMES AT THE SHORE
Lava flows into the sea from Hawaii, making new land. The Pacific has more volcanic islands than any other ocean.

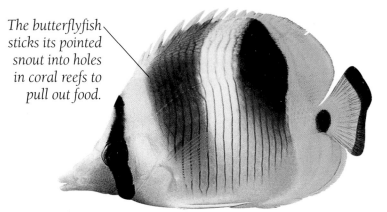

The butterflyfish sticks its pointed snout into holes in coral reefs to pull out food.

FLAT AND FINNY
The flat body of the butterflyfish is handy for swimming through tight spots in coral reefs.

ISLAND ERUPTION
As a huge plate carrying the Pacific floor moves over hot, liquid rock, called magma, the magma bursts through the plate, creating volcanoes. The Hawaiian islands were formed this way.

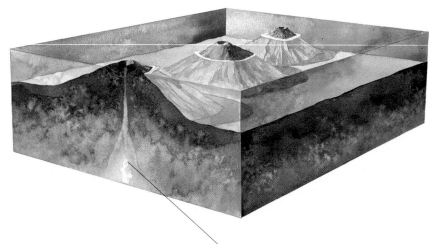

Hot magma punches through the moving Pacific Plate like a sewing needle, making a chain of islands.

GIANT CLAM, GIANT REEF
A giant clam lives on the Great Barrier Reef. This system of reefs stretches 1,250 miles along Australia's east coast.

The sea dragon's "leaves" are actually flaps of skin.

PLANT OR ANIMAL?

The leafy sea dragon, a relative of the seahorse, looks like seaweed floating in Pacific waters.

SAVE MY HOME

The Hawaiian monk seal, one of the few seals to live in warm waters, is endangered because people have harmed its habitat.

Atlantic Ocean

Despite cold weather in its northern half, the S-shaped Atlantic sees heavy ship traffic between Europe and North America. The most dramatic feature of the Atlantic is hidden underwater. Called the Mid-Atlantic Ridge, it is a huge undersea mountain range that snakes through the middle of the ocean. The ridge marks the place where two plates of the Earth's surface are moving apart.

ICE MONSTER
As huge as mountains or as small as boulders, icebergs sail the chilly waters of the North Atlantic. Many of them have broken off the ice sheet that covers Greenland, the world's largest island.

The sandtiger stays afloat by gulping air into its stomach.

OPEN WIDE
The slow-moving sandtiger is one of many sharks that haunt Atlantic coasts. Like most sharks, it is not dangerous to humans. Instead, the ten-foot-long swimmer hunts smaller creatures and other sharks in the rich fishing grounds of the Atlantic.

A MORAY STORY
A moray eel looks a photographer in the eye from its hideaway in the Caribbean Sea. The Caribbean, southeast of the United States, is a warm part of the Atlantic known for its coral reefs.

MOUNTAINS NO ONE CLIMBS
The rugged mountains of the Mid-Atlantic Ridge rise up to two miles high, with their peaks about a mile below the ocean's surface. A deep rift, or crack, splits them down the middle for much of their length. Molten rock from inside the Earth wells up into this rift.

The rift of the Mid-Atlantic Ridge measures almost a mile from rim to floor.

Indian Ocean

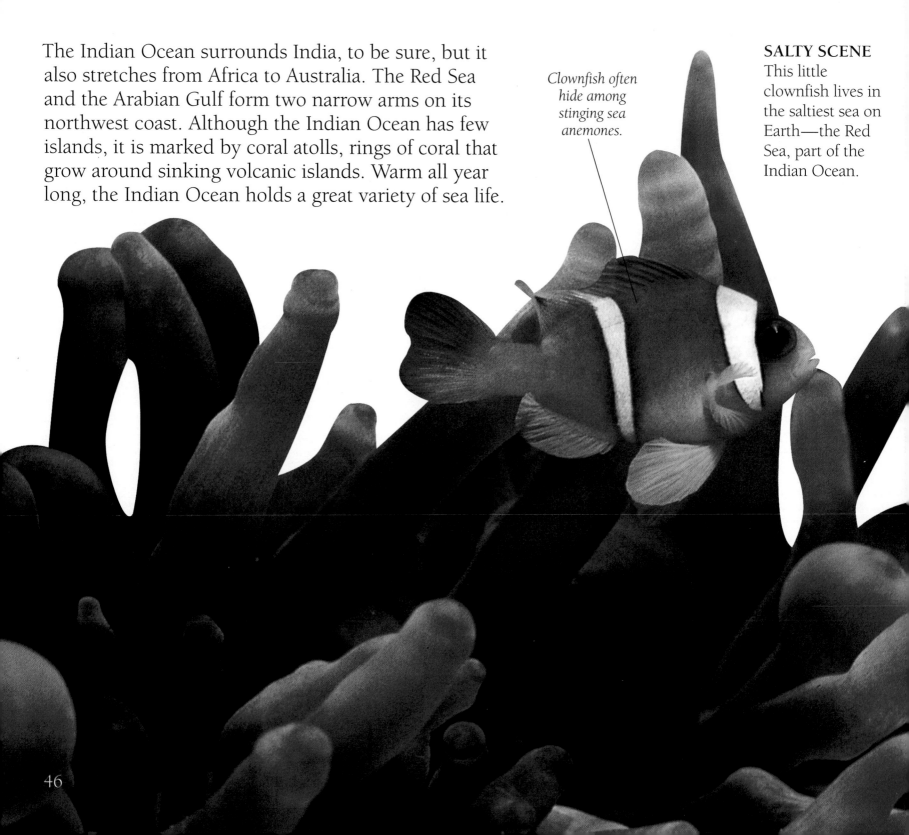

The Indian Ocean surrounds India, to be sure, but it also stretches from Africa to Australia. The Red Sea and the Arabian Gulf form two narrow arms on its northwest coast. Although the Indian Ocean has few islands, it is marked by coral atolls, rings of coral that grow around sinking volcanic islands. Warm all year long, the Indian Ocean holds a great variety of sea life.

Clownfish often hide among stinging sea anemones.

SALTY SCENE
This little clownfish lives in the saltiest sea on Earth—the Red Sea, part of the Indian Ocean.

WET AND WILD
Powerful winds sweep through palm trees near the Indian Ocean. This ocean is known for its changeable currents and intense seasonal rainfall.

LIVING FOSSIL
Scientists once thought that coelacanths (SEE-luh-kanths) died out with the dinosaurs. Now they know that a few of these six-foot rarities still live in the Indian Ocean.

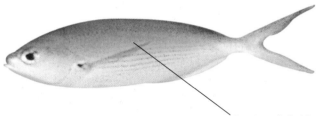

Their colors help fusilier fish blend into the blue depths of the sea.

THROUGH THE LOOKING GLASS
Blue-and-gold fusilier (FYOO-zeh-leer) fish form schools in the Indian Ocean.

TOUGH TIMES FOR CORAL
Coral reefs make for rough shorelines along the island of Sri Lanka. Coral in the Indian Ocean has suffered in recent years. Overbuilding, pollution, and overly warm waters are killing the algae that help the coral survive.

47

Arctic Ocean

There is no land at the North Pole—none for hundreds of miles. The north end of our planet is capped by the icy Arctic Ocean, the smallest and shallowest of all the oceans. Most of the water is covered by pack ice, which can be more than 160 feet thick in winter. The ice drifts on Arctic currents. In spite of the Arctic's frigid weather, many animals live around the ocean's coasts, including polar bears, seals, walruses, and whales.

ICE BEAR
The polar bear, one of the largest predators on land, may roam for thousands of miles across the Arctic ice.

Adult polar bears are more than five feet tall from paw to shoulder.

Most chitons are small—only two inches long at most.

ARCTIC ARMOR
You can always tell a chiton (KI-ton), by its eight overlapping plates. This one lives on Arctic shores, clinging to rocks.

SNOW SEAL
A young harp seal lounges on the Arctic ice. Arctic seals, protected by a thick layer of blubber, aren't threatened by the cold. They do face danger from hungry polar bears and, at times, from humans. Seals have been hunted for their meat, skin, and the oil from their blubber.

AN OCEAN IN MOTION
Arctic pack ice is always splitting up, moving about, colliding, and freezing again. In summer, it melts into separate floating chunks.

Only the male narwhal grows this long tooth.

USE A BIG TOOTHBRUSH
The narwhal, a small Arctic whale, is known for its "horn"—which is actually a long tooth. In medieval times people believed these teeth were unicorn horns.

49

Antarctic Waters

Three oceans—the Pacific, Atlantic, and Indian—meet around the continent of Antarctica. Despite the region's intense cold, Antarctic waters are rich in plant and animal life. Currents sweeping around the land and upward from the seafloor stir up nutrients that feed animals from plankton to whales.

The icefish survives in Antarctic waters because of a natural chemical in its blood that works like antifreeze.

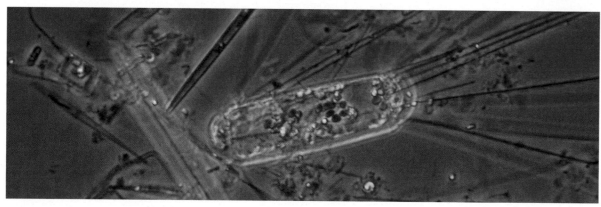

TINY SURVIVORS

More than a hundred kinds of algae, tiny plant-like organisms, live in the ocean near Antarctica. They can survive even when frozen into the ice in winter.

Waters of the Antarctic Convergence extend out from the continent of Antarctica for hundreds of miles.

A WATERY MIXING BOWL

Warm waters from the north meet cold waters from the south in a unique zone called the Antarctic Convergence.

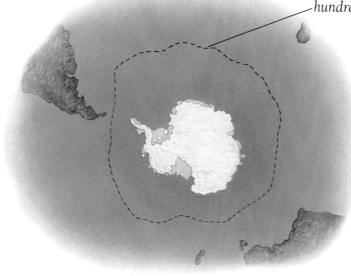

SOUTHERN SEAL

The Weddell seal of Antarctica lives farther south than any other mammal. It is a deep-diving champ, able to go as much as 1,900 feet underwater and to hold its breath for more than an hour.

HOW'S THE WEATHER?

A researcher releases a weather balloon above the ice-covered surface of the Bellingshausen Sea. Scientists from many countries come to Antarctica to study its weather and waters.

TOUGH BIRDS

Emperor penguins spend the winter in Antarctica. These hardy seabirds are the largest and heaviest of seven kinds of penguins that nest in Antarctica.

A thick layer of fat protects penguins from cold. The birds can live on their stored fat for months.

51

6 The Human Factor

In the past, people saw the ocean mainly as a route for travel or as a source of swimming food. Now we realize that we must take care of the ocean to keep planet Earth healthy. In addition to providing food for the world, the ocean helps give us clean air and helps keep us warm. Now people are trying to stop pollution from entering the world ocean.

MEASURING UP
Two workers at a U.S. marine sanctuary measure a young loggerhead turtle. All sea turtles are scarce. They are protected in United States waters.

CATCH OF THE DAY
A fishing boat in Iceland hauls in a day's catch. Fishing fleets around the world pull in more than a hundred million tons of fish and other creatures each year. Cod, haddock, sardines, and anchovies are among the most common fish caught for human and animal food.

POISONED WATER

Ocean pollution comes from many places. Some homes and factories dump wastewater directly into the sea. Rain washes chemicals such as weed-killers from lawns and farms into rivers—and eventually into the ocean.

DEADLY SPILL

When ships carrying petroleum accidentally spill their loads into the ocean, seabirds and mammals are among the first to be hurt. Sea otters often die when their fur is covered with oil.

Exploration Today

More than 95 percent of the ocean is unexplored. It is the last frontier on Earth. One of the main barriers to deep-sea exploration has been pressure. Human bodies are designed for air pressure—the weight of air—at the ocean's surface. Water is heavier than air. For every 33 feet a person dives, pressure doubles. Explorers must build special submarines, called submersibles, to take them into the deep sea.

When an ordinary Styrofoam cup is pulled deep underwater next to a submersible, pressure from tons of water crushes it into a tiny version of its former self.

GOING DOWN!
Scientists board the submersible *Alvin* as they prepare to dive into the undersea valley known as the Cayman Trench. *Alvin* can dive to 13,000 feet below the ocean's surface. *Alvin* was one of the subs used to film the wrecked remains of the *Titanic*.

HANGING AROUND
Two researchers suspended above Antarctic waters search for algae in the ice. Polar seas still hold many mysteries. Scientists want to know how life has managed to survive in such extreme conditions.

A SNUG FIT
Marine biologist Sylvia Earle climbs into her one-person submersible, *Newton,* which allows her to dive safely to one thousand feet.

Did You Know...

1 **THAT** coral can grow to look like all kinds of plants or animals? The coral at left is called mushroom coral. Each rounded piece is a single coral animal. Other kinds of corals look like boulders, deer's horns, and brains.

2 **THAT** whales can sing? The humpback whale, in particular, is known for its beautiful songs. The whales repeat phrases in the songs for hours or days. These marine mammals use the songs to communicate with each other.

3 **THAT** the biggest storm wave ever seen was 112 feet high—as tall as a ten-story building? The average ocean wave is only five to ten feet high.

4 **THAT** if ice did not float, there might be no life on Earth? If ice formed at the cold bottom of the sea, the ocean would freeze from the bottom up. There would be no ocean as we know it now.

5 **THAT** people in some countries actually take the salt out of seawater to use on their food? Sodium chloride, the chemical that makes up common table salt, dissolves in water. Sodium is what makes seawater taste salty.

6 **THAT** Australia's Great Barrier Reef is the world's largest coral reef? The reef is actually the largest thing in the world built by living creatures. Each piece of this huge structure was built by animals no bigger than your fingernail.

Glossary

ADAPT To change in order to fit in better with the environment.

BACTERIA Tiny, one-celled organisms found in the bodies of animals, organic matter, plants, soil, water, and air.

BALEEN A hard, comb-like material that grows from the upper jaws of baleen whales and is used for filtering food from water.

COAST The land next to the sea.

COLONY A group of the same kind of animals or plants living together.

CONTINENT One of the main divisions of land on Earth. The seven continents are Asia, Africa, North America, South America, Antarctica, Europe, and Australia.

CRUSTACEAN An animal with a hard shell and joined body and legs that lives mostly in water, such as a crab, a lobster, or a shrimp.

CURRENT A distinct stream of water that flows within an ocean.

ENDANGERED In danger of becoming extinct.

ESTUARY The place where a river meets the ocean's tides.

HABITAT An organism's natural home, such as the ocean, a river, a forest, or a desert.

INTERTIDAL Between high-water and low-water levels along a coast.

LAVA Hot, melted rock that flows onto the Earth's surface.

MARINE Belonging to the ocean.

MINERAL A nonliving element, such as iron, sodium, or calcium, that is found in nature.

MOLLUSK A soft-bodied animal that lacks a backbone and is not divided into segments. A mollusk, such as a clam, is sometimes covered with a hard shell.

ORGANISM Any single form of life.

PETROLEUM An oily, dark liquid that is taken out of the earth and used for fuel.

POLLUTION Wastes and poisons released into air, water, or on land.

PREDATOR An animal that hunts and kills other animals for food.

REEF A ridge of rock, coral, or sand near the surface of the water.

SANCTUARY A place where animals can be protected from danger.

SCAVENGE To gather something usable, such as food, from material that has died or been thrown away.

TRENCH A long, narrow valley under the sea.

Index

Boldface indicates illustrations.

Albatross 28, 29, **29**
Algae 47, 50, **50**
Anchovies 52
Anemones: tube 36, **36**
Anglerfish **5**, 34, **34**
Antarctic Convergence 50, **50**
Antarctic waters 40, 50–51; exploration 55
Arctic Ocean 40, 48–49
Atlantic Ocean 40, 44–45

Bacteria: anglerfish 34; deep-ocean vents 38–39
Baleen, whale 15
Barnacles: goose 18, **18**
Barracuda 25, **25**, 26, **27**
Basket star 37, **37**
Blubber 30, 48
Butterflyfish 42, **42**

Cayman Trench 54
Chiton 48, **48**
Clams: deep ocean 38; giant 42, **42**
Clownfish 47, **47**
Coasts 16–25; coral reefs 24–25; kelp 20–21; mud 22–23; tides 18–19; *see also* Shore
Cod 52
Coelacanths 46, **46**
Coral: Great Barrier Reef 42, **42**; growth 24; Indian Ocean 46, **46**, 47; mushroom 56, **56**; poisonous animals 16, **16**; polyps 16, **17**, 24, **24**; reefs 24–25
Crabs: fiddler 22, **22**; giant spider 37, **37**; hermit 18, **18**
Currents: ocean 10–11
Cutlass fish 36, **36**

Deep ocean 13, **13**, 34–39
Dolphins 30, **30**

Earle, Sylvia 55, **55**
Exploration: ocean 54–55

Food web 14–15
Fusilier fish 46, **46**

Giant tube worms 38, **38–39**
Great Barrier Reef, Australia 42, **42**, 57, **57**
Grouper 32, **32**
Gulf Stream 10, **11**

Haddocks 52
Hatchetfish 34, **34**
Hawaii: volcanoes 42, **42**
Horseshoe crab 22, **22**

Ice 57, **57**; Arctic Ocean 48, 49, **49**
Icebergs 44, **44**
Icefish 50, **50**
Indian Ocean 40, 46–47

Jellyfish **4**, 33, **33**

Kelp 20–21; giant 21, **21**
Krill 15, **15**, 26, **26**; *see also* Plankton

Lava: pillow 36, **36**; *see also* Magma
Lionfish **4**, 24, **24**
Lobsters: blind 38

Magma 42, **42**; *see also* Lava
Manatees **4**, 14
Mariana Trench 42
Mid-Atlantic Ridge 44, 45, **45**
Moray eel 45, **45**

Mudskipper 22, **22**

Newton (submersible) 55, **55**
Nudibranchs 18, **18**

Ocean: facts 6; names 40–41, **40–41**; under 8–9
Ocean floor 12, **12**, 36–37
Octopus, blue-ringed 16, **16**

Pacific Ocean 40, 42–43
Penguins, emperor 51, **51**
Plankton 15, 26, **26**; purple 35, **35**; *see also* Krill
Plants 14; coral reefs 24–25; kelp 20–21; ocean zones 12–13, **12–13**
Plates: black smokers 38, **38**; Earth's 8, **8**; magma 42, **42**; pillow lava 36, **36**
Polar bears 48, **48**
Pollution 52, 53, **53**
Pressure, water 54, **54**

Red Sea **4**, 8, **8–9**, 46

Sailfish 32, **32**
Salt 6, 56, **56**
Sand 17, **17**
Sardines 52
Scorpionfish 36, **36**
Sea cucumbers 22
Sea dragon: leafy **5**, 43, **43**
Sea nettles 33
Sea otters 53, **53**; food 15, **15**
Sea slugs 18, **18**
Sea stars 19, **19**; crown-of-thorns 25, **25**
Sea turtles 26, **26**, 52, **52**
Sea urchins 20; food 14
Seabirds 28–29
Seals: gray 30, **30**; harbor 16,

16, 20, **20**; Hawaiian monk 43, **43**; harp 48, **48**; Weddell 50, **50**
Sharks: sand tiger 45, **45**; reef 32, **32**
Shore 12–13, **12–13**; *see also* Coasts
Shrimp 35, **35**
Silversides 25, **25**
Spoonbills, roseate 23, **23**
Squids: giant 35, **35**
Starfish *see* Sea stars
Submersibles **5**, 54, **54**, 55, **55**
Sunlit waters 12, 13, **12–13**, 26, 26–33
Sweeper fish 33, **33**

Tides 18–19; Bay of Fundy, Canada 19, **19**; tidal pool 18, **18**
Titanic (ship) 54
Trenches 8, **8**; Cayman 54; Mariana 42
Tripod fish 36, **36**
Tuna 32, **32**
Twilight zone 13, **13**

Vents 38–39
Viperfish 34, **34**
Volcanoes: Pacific Ocean 42, **42**

Waves **4**, 8, **8–9**; highest 57, **57**
Weather balloon 51, **51**
Whales: ancestor 30, **30**; beluga 30, **31**; blue 15, **15**; humpback 30, **30**; narwhal 49, **49**; songs 56, **56**

Zones: ocean **12–13**

Credits

sea anemone

Published by

The National Geographic Society
John M. Fahey, Jr., *President*
 and Chief Executive Officer
Gilbert M. Grosvenor,
 Chairman of the Board
Nina D. Hoffman,
 Senior Vice President
William R. Gray, *Vice President and Director, Book Division*

Staff for this Book

Barbara Brownell, *Director of Continuities*
Marianne R. Koszorus, *Senior Art Director*
Toni Eugene, *Editor*
Alexandra Littlehales, *Art Director*
Susan V. Kelly, *Illustrations Editor*
Patricia Daniels, *Researcher*
Sharon Kocsis Berry, *Illustrations Assistant*
Deborah Patton, *Indexer*
Mark A. Caraluzzi, *Director of Direct Response Marketing*
Heidi Vincent, *Marketing Manager*
Vincent P. Ryan, *Manufacturing Manager*
Lewis R. Bassford, *Production Project Manager*

Acknowledgments

We are grateful for the assistance of marine biologist
Sylvia A. Earle, Ph.D., *Scientific Consultant.*

Illustrations Credits

COVER: Beverly Factor/Adventure Photo & Film
Interior photographs from National Geographic Image Sales
Front Matter: 1 George Grallings; 2-3 George Mobley; 4 (top to bottom) Nick Caloyianis; Bianca Lavies; George Grallings; Sisse Brimberg; 5 (top to bottom) Darlyne A. Murawski; David Doubilet; Emory Kristof; 6-7 (art) The Studio of Wood Ronsaville Harlin; 8 (art)The Studio of Wood Ronsaville Harlin; 8-9 Nick Caloyianis; 10 Jen & Des Bartlett; 10 (art) Robert Cremins; 11 NOAA.
Ocean Zones: 12-13 (art) The Studio of Wood Ronsaville Harlin; 14 (top) George Grallings; (bottom) Bianca Lavies; 15 Sisse Brimberg; 15 (art, both) Robert Cremins.
Coasts: 16 (art) The Studio of Wood Ronsaville Harlin; 16 Richard Olsenius; 17 (top left) Marc Moritsch; (top right) Medford Taylor; (bottom) Nick Caloyianis; 18 (top) Robert Sisson; (bottom) George Grallings; 18 (art) Robert Cremins; 19 Raymond Gehman; 19 (art) Carol Schwartz; 20 David Doubilet; 21 Bates Littlehales; 21 (art) The Studio of Wood Ronsaville Harlin; 22 (left) Luis Marden; (right) Bates Littlehales; 22 (art) Robert Cremins; 23 (top) Raymond Gehman; (bottom) Chris Johns; 24 (both) George Grallings; 24 (art) Carol Schwartz; 25 Nick Caloyianis; 25 (art) Robert Cremins.
The Sunlit Waters: 26 (top) George Mobley; (center) Bill Curtsinger; (bottom) Maria Stenzel; 26 (art) Robert Cremins; 27 Bill Curtsinger; 28 Nick Caloyianis; 29 (top) Richard Olsenius; (bottom) Jen & Des Bartlett; 29 (art) Robert Cremins; 30 (top) David Doubilet; (center) John Eastcott & Yva Momatiuk; (bottom) Bill Curtsinger; 30 (art) Robert Cremins; 31 Nick Caloyianis; 32 (both) Bill Curtsinger; 32 (art) Robert Cremins; 33 (top) Sisse Brimberg; (bottom) David Doubilet.
The Open Ocean: 34 (left) Darlyne A. Murawski; (top & bottom) Paul Zahl; 35 (both) Paul Zahl; 35 (art) Robert Cremins; 36 (both) Emory Kristof; 36 (art top) Robert Cremins; (art bottom) Carol Schwartz; 37 (both) Emory Kristof; 38 (art) The Studio of Wood Ronsaville Harlin; 38-39 Emory Kristof.
The Ocean Tour: 40-41 (art) The Studio of Wood Ronsaville Harlin; 42 (left) George Grallings; (right) Chris Johns; (bottom) Bates Littlehales; 42 (art) The Studio of Wood Ronsaville Harlin; 43 (top) David Doubilet; (bottom) Bill Curtsinger; 44 Richard Olsenius; 45 (both) George Grallings; 45 (art) Carol Schwartz; 46 (left) Priit J. Vesilind; (right) David Doubilet; 46 (art) The Studio of Wood Ronsaville Harlin; 47 (top) Mattias Klum; (bottom) James Stanfield; 48 (top) Lowell Georgia; (bottom) George Mobley; 49 (top) Otis Imboden; (bottom) Nick Caloyianis; 49 (art) The Studio of Wood Ronsaville Harlin; 50 (left) Maria Stenzel; (right) Jen & Des Bartlett; 50 (art top) Robert Cremins; (art bottom) Carol Schwartz; 51 (both) Maria Stenzel.
The Human Factor: 52 (top) Wolcott Henry; (bottom) James Sugar; 53 Bates Littlehales; 53 (art) Robert Cremins; 54 Emory Kristof; 54 (art) Robert Cremins; 55 (left) Maria Stenzel; (right) Nate O. Johnson.
Backmatter: 56 George Grallings; 56 (art both) Robert Cremins; 57 (top) Maria Stenzel; (bottom) Gordon Gahan; 57 (art) Carol Schwartz; 60 George Grallings.

COVER: Viewed from below, a sea turtle cruises blue Pacific waters off Kona Island, Hawaii.

Composition for this book by the National Geographic Society Book Division. Printed and bound by R.R. Donnelley & Sons Company, Willard, Ohio. Color separations by Graphic Arts Services, Nashville, Tennessee. Case cover printed by Inland Press, Menomonee Falls, Wisconsin.

Library of Congress CIP Data
Daniels, Patricia.
 Ocean / Patricia Daniels.
  p. cm. -- (National Geographic nature library)
 Summary: Describes the world's oceans, their physical features, and the animals and plants that live in them.
 ISBN 0-7922-7545-4
 1. Ocean--Juvenile literature. 2. Oceanography--Juvenile literature.
[1. Ocean.] I. Title.
 II. Series.
GC21.5 .D36 1999
 551.46--dc21
 99-044330